Amrit Sarkar

Human Activity and Behavior Recognition in Videos. A Brief Review

GRIN Verlag

Bibliografische Information der Deutschen Nationalbibliothek:

Die Deutsche Bibliothek verzeichnet diese Publikation in der Deutschen National-
bibliografie; detaillierte bibliografische Daten sind im Internet über http://dnb.d-
nb.de/ abrufbar.

Imprint:

Copyright © 2014 GRIN Verlag GmbH
Druck und Bindung: Books on Demand GmbH, Norderstedt Germany
ISBN: 978-3-656-69126-6

This book at GRIN:

http://www.grin.com/en/e-book/276054/human-activity-and-behavior-recognition-
in-videos-a-brief-review

GRIN - Your knowledge has value

Der GRIN Verlag publiziert seit 1998 wissenschaftliche Arbeiten von Studenten, Hochschullehrern und anderen Akademikern als eBook und gedrucktes Buch. Die Verlagswebsite www.grin.com ist die ideale Plattform zur Veröffentlichung von Hausarbeiten, Abschlussarbeiten, wissenschaftlichen Aufsätzen, Dissertationen und Fachbüchern.

Visit us on the internet:

http://www.grin.com/

http://www.facebook.com/grincom

http://www.twitter.com/grin_com

Human Activity and Behavior Recognition in Videos: A Brief Review

Amrit Sarkar

B.Tech, Computer Engineering
National Institute of Technology, Kurukshetra

Abstract— **Understanding human activity and behavior, especially real-time understanding of human activity and behavior in video streams is presently one of the most active areas of research in Computer Vision and Artificial Intelligence. Its purpose is to automatically detect, track and describe human activities in a sequence of image frames. Challenges in this topic of research are numerous and sometimes very difficult to work out. This paper presents a brief review over the overall process of Human Activity and Behavior Recognition both real time and non-real time, and some of the applications present in current world. The main purpose of this survey is to extensively identify some of the existing methods, critically analyze it and acknowledge the work done by researchers in this field so far.**

I. INTRODUCTION

As an active research topic in computer vision, visual components in dynamic scenes attempts to detect, recognize and track certain objects from image sequences, and more generally to understand and describe object behaviors.

The aim is to develop intelligent video processing systems which can semantically detect human activities and their behavior. We need intelligent visual surveillance to replace the traditional passive video surveillance that is proving ineffective as the number of cameras exceeds the capability of human operators to monitor them. In short, the goal of visual surveillance is not only to put cameras in the place of human eyes, but also to accomplish the entire surveillance task as automatically as possible. Significant technological advances in hardware and communication protocols are also facilitating new services, such as real time collection of statistics on group sports and annotation of videos for event detection and retrieval.

In a surveillance environment, the automatic detection of abnormal activities can be used to alert the related authority of potential criminal or dangerous behaviors, such as automatic reporting of a person with a bag loitering at an airport or station. Similarly, in a healthcare system, the activity recognition can help the rehabilitation of patients, such as the automatic recognition of patient's action to facilitate the rehabilitation processes Furthermore in Sports, viewers demand a more augmented visual display of game for better understanding while coaches and players want to extract relevant information for performance and tactics upgrade. With increasing demand of precision in decisions by referees/umpires/officials, intelligent video processing systems are now an eye-popper in sports.

A number of processing steps are necessary to analyze the scene at different levels of abstraction, starting from the behaviors of objects of interest. The first step consists of detecting and tracking subject(s) of interest to generate motion descriptions (e.g., motion trajectory or combination of local motions), which are then processed to identify actions or interactions. Depending on the quality of the camera view, position information can be complemented by other descriptors, such as the trajectories of body joints or head pose changes.

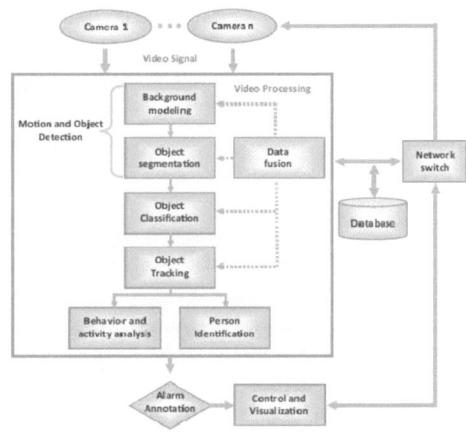

A broad picture of an Intelligent Video Surveillance System: Figure: taken from [1].

The paper is organized as follows. In Section II, a brief review of Motion and Object Segmentation is presented. Section III presents a review of Object Classification and Tracking, whereas Section IV discusses Extraction of features and Motion information. Section V gives a review of various Activity and Behavior Recognition Methods and Techniques. Some Recent Work and Applications are briefed in Section VI and finally, Section VII concludes the paper with an overview on future trends and open research issues.

II. MOTION AND OBJECT SEGMENTATION

Most visual surveillance systems start with motion detection. Motion detection methods locate connected regions of pixels that represent the moving objects within a particular scene; Different approaches include frame-to-frame difference; background subtraction [2] and motion analysis using optical flow techniques [3] are used. Motion detection aims at segmenting regions corresponding to moving objects from the rest of an image. The motion and object detection process usually involves background modeling [4] and motion segmentation [5]. Subsequent processes such as object classification, tracking, and activity and behavior recognition are greatly dependent on it.

Original Video Frame   Segmentation   Detected Foreground

Humans are segmented out as points of interest in the Fig.

Based on the mobility of the camera, the object segmentation task can be divided into two categories, the static camera segmentation and the moving camera segmentation.

1) Static Camera

In static camera segmentation, the camera is fixed in a specific position and angle. Since the background never moves, it is natural to build a background model in advance, so that the foreground object can be segmented from the image of the background model.

a) Background Subtraction:

The most common method for static camera segmentation is background subtraction [2] [6] [7] due to its simplicity and efficiency. The background model contains only the stationary background scene without any foreground object, and any image change is assumed to be caused only by moving objects. Hence the foreground object can be obtained by subtracting the current image of the background image, followed by a calculating a threshold for segmentation masking. Background subtraction methods store an estimate of the static scene, accumulated over a period of observation; the objects which are not part of static scene are taken as foreground objects.

b) Gaussian Mixture Model (GMM)

Instead of the simple one-Gaussian per pixel background modeling [8] [9], the pixel values at location (x,y) can be modeled as a mixture of Gaussians (MoG), to accommodate different background scenarios. The Gaussian mixture model (GMM) [10] [7] [11] has been extensively applied in many fields to allow the adaptation to the multi-modal environments. Generally, GMM is learned by the expectation maximization (EM) algorithm [12] [13]. The higher the probability of a pixel value in the GMM, the more likely the pixel belongs to the background. Therefore, for the image sequence, a pixel of the image is classified to belong to the foreground object if the probability of the pixel value is less than a threshold which is predefined.

2) Multiple Camera

Moving camera segmentation is much more challenging than static camera segmentation. Apart from the motion of each foreground moving object, the motion of the background comes into play.

a) Temporal Difference

The most common method for moving camera segmentation is the temporal difference [1] [14] between consecutive frames. Unlike static camera segmentation, where the background is comparably stable, the background is changing along time for moving camera; therefore, it is not appropriate to build a background model in advance. Instead, the moving object is detected by taking the difference of consecutive image frames t-1 and t. However, the motion of the camera and the motion of the object are mixed in the moving camera [10]. Hence, the motion of camera is estimated first. It makes use of the pixel-wise difference between two to three consecutive frames in an image sequence to extract moving regions. Temporal differencing is very fast and adaptive to dynamic environments, but generally does a poor job of extracting all the relevant pixels, e.g., there may be holes left inside moving entities.

b) Optical Flow

Optical flow [14] [15] [16] is a vector-based approach that estimates motion in video by matching points on objects over multiple frames. Optical flow methods are very common for assessing motion from a set of images. However, most optical flow methods are computationally complex, sensitive to noise, and would require specialized hardware for real-time applications.

III. OBJECT CLASSIFICATION AND TRACKING

Different moving regions may correspond to different moving objects in corresponding scenes. To further track objects and analyze their behaviors, it is essential to correctly classify moving objects. For instance, the moving objects are humans, vehicles, or objects of interest of an investigated application. Object classification can be considered as a standard pattern recognition task. There are two main categories of approaches for classifying moving objects: shape-based classification [17] [19] and motion-based classification [18] [19], Different descriptions of shape information of motion regions such as points, boxes, silhouettes and blobs are available for classifying moving objects. In general, human motion exhibits a periodic property, so this has been used as a strong urge for classification of moving objects also.

Classification or pattern recognition algorithms [20] are mathematical tools for detecting similarities between members of a collection of cases. The output of a classification algorithm is used to classify new cases into subset. We briefly introduce three classifications techniques used in current world:

1) Linear Discriminant Analysis:

Linear discriminant analysis is a technique used to identify structures and possible organizations of the data into meaningful groups [21]. Any given set of features is related to a case is taken as a multidimensional space where each case is represented as a point with distinct coordinates. A group can be stated as any subset of the points which is internally well connected and externally poorly connected. A forward stepwise procedure is adopted which begins by selecting the individual feature which provides the greatest discrimination in terms of variation. The procedure then pairs this first feature with each of the remaining features, one at a time, to locate the combination which produces the greatest discrimination. The feature which contributed to the best pair is selected.

2) Classification Trees:

With the development of machine learning theory, many machine learning systems have been developed for constructing classification trees from a training set of examples. A decision tree is constructed using a top-down, divide-and-conquer approach: for the two group problem, select a feature, divide the training set into subsets characterized by the possible values of the feature, and follow the same procedure recursively with each subset until no subset contains cases from both groups [24].

3) Neural Network Approaches:

Neural networks are adaptive systems that learn from examples. They are networks of simple processing nodes that are usually interconnected. One popular neural network architecture for pattern recognition and classification is the back-propagation model [22] [23]. It has a three-layer, feed-forward structure. Both supervised and unsupervised classification can be done through neural networks. The more detailed explanation of the structure and algorithm is provided in Section V.

After motion detection, surveillance systems generally track moving objects from one frame to another in an image sequence. Region based tracking algorithms [25][26] track objects according to features extracted from image regions corresponding to the moving objects. While Contour [27] can provide an accurate description of shape for the objects such as hands, head and human by representing their outlines as bounding contours. Feature based tracking methods [28] use the features such as distinguishable points or lines on the object to determine the tracking task. It includes feature extraction and feature matching. Model-based tracking algorithms [14] [31] track objects by matching projected object model. The models are usually constructed offline with manual measurement, CAD tools or computer vision techniques. Other than these, hybrid based tracking is the combination of the stated tracking methods. 3D tracking includes volumetric models [29] and graphical models [30] to track the desired object.

Useful mathematical tools for tracking include the Kalman filter:
A Kalman Filter [7] [32] [33] is a set of mathematical equations to minimize the mean square errors (MSEs) and estimate the state of the dynamic process. Two phases in a Kalman filter are performed recursively, i.e., the estimation (prediction) of the process state and the update of the process

state by the measurement. The main limitation of Kalman filter is that it needs good foreground segmentation; hence, it has little ability to handle the occlusion.

Fig: Players being tracked in a football match. Rectangular box shows the marking.

A model-based tracking approach using Kalman Filters in Road Traffic scenes is given in [37].

Other than Kalman Filters, Condensation algorithm [35] [36], Dynamic Bayesian network etc. [37] [38] are other tools for tracking.

IV. EXTRACTION OF FEATURES AND MOTION INFORAMTION

The first important step in motion-based recognition is the extraction of motion information from a sequence of images. These features are the input sequences for the statistical, syntactic and probabilistic models for activity and behavior recognition. There are generally three methods for extracting motion information from a sequence of images [1]: Optical flow, Trajectory-based features, and Region-based features.

1) Optical Flow Features

Optical flows recover image motion at each pixel from spatio-temporal image on the basis of brightness variations. It is the velocity field, which warps one image into another (usually very similar) image. Several methods have been developed; however, a robust method is yet to be discovered [39].

2) Trajectory-based Features

Trajectories are very popular as they are relatively simple to extract and model. These trajectories are derived from the locations of the particular points on an arbitrary object in a fixed span of time [40]. Trajectories of motions are generated from sequence of images in which tokens in each frame of a scene is detected and correspondence of such tokens from one frame to another. These tokens should be precise and stable so that distinction and tracking can be made easily. Tokens can be a corner, region, a body part or a point of interest. Several proposed solutions [40] [41] for human actions modeling and recognition using the trajectory-based features approach are present. Steiger et al. [42] adopt global and local reference points to characterize motion information. In a general trajectory-based features approach, an arbitrary changing number of objects are tracked initially. Temporal trajectories are formed, which describe the motion paths of these objects, are derived from history statistics. Then, characteristic motion patterns are determined, by clustering the trajectories into prototype curves. Finally, motion recognition is then derived by tracking the position within these prototype curves based on the same method used for the object tracking.

3) Region or Image-based Features

In some cases, for some objects or motions, precise motion information is not desired instead a general idea of the frame is helpful. Features generated from the use of information over a relatively large region or over the whole image are referred as region-based features. An effective approach been given in [43].

Fudan University and City University of Hong Kong [44] detects violent scenes in video which has potential for several applications. Dense local patch trajectories are first extracted, based on which motion representations are generated by exploring relative locations and motions between trajectory pairs. While Spatial-Temporal Interest Points (STIP) captures a space-time volume in which video pixel values have large variations in both space and time. Concept based features are also taken into account.

V. ACTIVITY RECOGNITION METHODS AND TECHNIQUES

After Object Detection, Classification and Tracking and Feature extraction, depending on the amount of prior knowledge and human involvement in the learning process, we may broadly categorize the research in activity and behavior detection, especially abnormal behavior detection, as supervised, unsupervised, and semi-supervised [45]. We can say a behavior is abnormal in a scene, when it is not supposed to happen at that point of time in that particular scene.

Supervised methods build models of normal or/and abnormal behavior based on the labeled data. Video segments that do not fit the models are "flagged off" as abnormal. This modeling approach for unusual events' detection is good only if these abnormal events are well defined and there are enough training data. While unsupervised methods learn the normal and abnormal patterns from the statistical properties of the observed data. Isolated clusters identified as anomalies. Semi-supervised approaches fall in-between the first two. They learn a model of usual/unusual events using partially labeled data. [45]

The k-means algorithm is used to cluster features. The training and learning process for anomaly detection involves grouping similar image/video clip descriptors together and create a finite number of clusters that have unique cluster structures and possible semantic meaning. Local features are extracted and clustered using low-level abstraction to describe activities. Objects that are not located in the main clusters of a set of data are regarded as anomalies/abnormal behaviors.

In [46], it was pointed out that different clustering algorithms can be characterized based on the following factors:

1) Normal data instances belong to a cluster in the data, while anomalies either do not belong to any cluster
2) Normal data instances lie close to their closest cluster centroid, while anomalies are far away from their closest cluster centroid
3) Normal data instances belong to large clusters, while abnormal data belong to sparse clusters.

In the remaining part of the Section, brief review of some of the existing activity and detection techniques; both supervised and unsupervised; is given.

A. DYNAMIC TIME WARPING:

DTW [47] [48] is one of dynamic programming algorithms to measure similarity (distance) between two sequences, i.e., one kind of template matching algorithms [49]. In [48] DTW is used to recognize various human activities such as waving, punching and clapping. Bobick et al. [50] use DTW to match a test sequence to a deterministic sequence of states to recognize human gestures. Even if the time scale between a test sequence and a reference sequence is inconsistent, DTW can still successfully establish matching as long as the time ordering constraints hold.

It has the advantage of conceptual simplicity and robust performance, and has been used recently in the matching of human movement patterns [50] [51] but it might need extensive templates for various situations, resulting in high computation cost to match with these extensive templates [10].

B. FINITE-STATE MACHINE (FSM):

The most important feature of a FSM [14] [52] is its state-transition function. The states are used to decide which reference sequence matches with the test sequence. Wilson et al. [53] analyze the explicit structure of natural gestures where the structure is implemented by an equivalent of a FSM but with no learning involved. State machine representations of behaviors have also been employed in higher level description.

It is a very efficient, accurate and robust algorithm to detect checkout primitive or basic activities [54] [55]. Experimental results [54] [55] show that this new approach, although simpler and more intuitive, outperforms more sophisticated machine learning-based techniques both in detecting primitives and more complex checkout activities.

C. HIDDEN MARKOV MODEL:

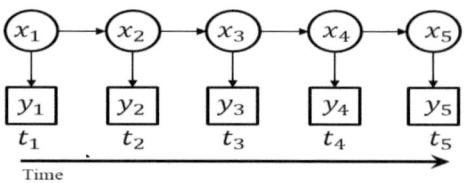

HMM frame [7]: Rectangular boxes depict the Observed states while Circular blobs hidden states

An HMM [50] [56] is specified by three terms. The first term is the initial probability of hidden states. The second term is the transition matrix, which specifies a transition probability from one hidden state to another hidden state. The third term is the observation matrix, which specifies the probability of the observed symbol given a hidden state.

A HMM is a kind of stochastic state machines [6] [57]. HMMs use hidden states that correspond to different phases in the performance of an action. The use of HMMs consists of two stages: training and classification. In the training stage, the number of states of a HMM must be specified, and the corresponding state transition and output probabilities are optimized in order that the generated symbols can correspond to the observed image features of the examples within a specific movement class. In the matching stage, the probability with which a particular HMM generates the test symbol sequence corresponding to the observed image features is computed.

HMMs can be used to solve three problems as given in [39]:
1) To evaluate the probability of a sequence of observed events given a specific model;
2) To determine the most likely evolution of an abnormal activity (state sequence) that is represented by the HMM
3) To estimate HMM parameters that produce the best representation of the most likely state sequence.

American Sign Language [58] has been recognized through HMMs. HMMs generally outperform DTW for undivided time series data, and are therefore extensively applied to behaviour understanding. More applications of HMMs are given in Section VI.

However, the Markovian framework makes strong restrictive assumptions about the system generating the signal, that it is a single process having a small number of states and an extremely limited state memory. Coupled HMM [59] overcome these limitations by offering a way to model multiple interacting processes without running afoul of the Markov condition. CHMMs couple HMMs with temporal, asymmetric conditional probabilities. In addition, CHMMs are far less sensitive to initial conditions than conventional HMMs, e.g., they are more reliable.

D. DYNAMIC BAYESIAN NETWORKS:

Dynamic Bayesian Network [7] [45] [61] is Bayesian network that represents sequences of variables and is considered as a generalization of the Hidden Markov Models. It fulfills the limitation of HMM of not being able to handle more than 2 processes at a time efficiently [57]. In this architecture, the low level Bayesian network estimates the human body part poses, and the high level Bayesian network estimates the overall body poses. A hierarchical framework is used to represent multiple event levels from the body-part level, to the multiple-bodies level, and finally, to the video sequence level. A good work has been done for handling occlusions that occur during human interactions through Bayesian network inference [60].

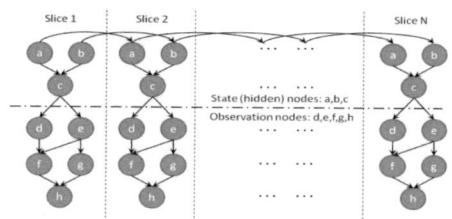

An example of a dynamic Bayesian network (DBN) [7] is unrolled in time axis with state (hidden) notes and observation nodes

E. ARTIFICIAL NEURAL NETWORKS:

Artificial neural networks (ANNs) [62] [63] are a computational structure that is inspired by observed process in natural networks of biological neurons in the brain. It consists of simple computational units called neurons, which are highly interconnected. Natural neurons receive signals through synapses located on the dendrites or membrane of the neuron. When the signals received are strong enough (surpass a certain threshold), the neuron is activated and emits a signal though the axon. This signal might be sent to another synapse, and might activate other neurons.

ANNs are constructed with layers of units, and thus are termed multilayer ANNs. ANNs [64] basically consist of inputs at first layer, which are multiplied by weights, and then computed by a mathematical function which determines the activation of the neuron. Another function computes the output of the artificial neuron which is done in the last layer. All the other layers are hidden layers in the ANN structure. ANNs combine artificial neurons in order to process information. The higher a weight of an artificial neuron is, the stronger the input which is multiplied by it.

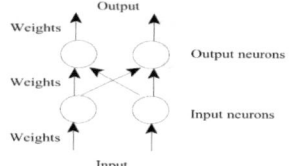

Diagrammatic representation of working of neurons in ANNs [62]

Depending on the weights, the computation of the neuron will be different. By adjusting the weights of an artificial neuron we can obtain the output we want for specific inputs. Algorithms are proposed in order to obtain the desired output from the network. This process of adjusting the weights is called learning or training. Both supervised and unsupervised learning is possible through ANNs.

The most widely used learning algorithm in an ANN is the Back-propagation algorithm [65] [66], which uses the data to adjust the network's weights and thresholds so as to minimize the error in its predictions on the training set. There are various types of ANNs like Multilayered Perceptron [67], Radial Basis Function [68] and Kohonen networks [64].

A neural network was used to understand and learn the background knowledge of videos by Alberto et al. [69]. The combination of the image retrieval approach with the neural network approach efficiently solves the problem of the apparent motion of a static background in a mobile camera, by moving from the old concept of the classical "frame difference" paradigm to raise an alarm to the new "known scene"- no alarm & "unknown scene"- alarm paradigm, where the meaning of scene is related to spatial-temporal events.

F. NONDETERMINISTIC-FINITE-STATE AUTOMATON (NFA):

A NFA or nondeterministic finite state machine [70] [71] is a finite state machine where for each pair of state and input symbols, there may be several possible next states. This distinguishes it from the deterministic finite automaton (DFA), where the next possible state is determined uniquely [14]. One may construct an equivalent DFA of a NFA, and vice-versa. Hence it can readily work like a FSM.

G. SYNTACTIC TECHNIQUES:

Semantic event models do well to capture the structure of the event, however they are unable to capture uncertainty and often are less efficient in the event recognition phase. Recently the grammatical approach has been used for visual behavior recognition [72]. Brand et al. [73] have used a simple non-probabilistic grammar to recognize sequences of discrete behaviors.

The idea is to divide the recognition problem into two levels [57]. The lower level is performed using probabilistic behavior models, such as HMMs, to output possible low-level temporal features. These outputs provide the input stream for a stochastic context-free parser [74] [75]. The grammar and parser provide longer range temporal constraints, removes confusion of uncertain low-level detection, and allow the inclusion of a priori knowledge [76] about the structure of temporal behaviors in a given domain [70].

Petri Nets were defined by Carl Adam Petri [77] as a mathematical tool for describing relations between conditions and events. Petri Nets are particularly useful to model and visualize behaviors such as sequencing, concurrency, synchronization and resource sharing. Piotr et al.[78] propose to apply a temporal logic formalism to specify event scenarios and further to translate them to Petri net structures; Fuzzy Semantic Petri Nets is introduced; that interprets a data resulting from a tracking algorithm, represents it as a content of a fuzzy ontology and detects event occurrences with a FSPN interpreter. But Petri Nets suffer from the disadvantage of having to manually describe the model structure.

Other than these techniques: Time-delay neural network (TDNN) [79] [80], Physical-Parameter-Based Methods and others are gaining appreciations which are categorized on the basis of their nature.

VI. SOME RECENT WORK AND APPLICATIONS

A review of overall framework of an Intelligent Video Surveillance System has been discussed so far in this paper. In this Section, some recent techniques, methods and applications proposed, experimented and introduced in this field are presented, which are broadly categorized into Surveillance (security purposes), Healthcare and Sports.

A. SURVEILLANCE

The frequency, severity and sophistication of security breaches and attacks are rising at a considerable rate. Security is an issue for a lot of businesses, especially if one owns a restaurant or a retail shop. Robbery, theft, shoplifting and terrorist attacks are increasing around the world. Because of this, businesses and governments are increasing their adoption of video surveillance cameras. In fact, many governments have passed strong regulations to install video surveillance cameras in places such as hospitals, airports, hotels, malls and transit stations. With advancement in technology and limitation of man-power availability in video surveillances, various applications in intelligent video surveillances has come up in security field like in Loitering, Crowd Handling, Face Detection, Posture and Gait Estimation and other crime-related activities. In the following part of this Section, some of the applications and novel techniques in surveillance fields are categorized and presented with brief summary.

1) Crowd Count and Behavior Detection:

Accurate people detection and their respective behavior can increase surveillance and management efficiency in public and private transportation and gathering by marking areas and semantically identifying the nature with high congestion or signaling areas which requires more attention than others. Estimation of a crowd in a ticketing systems or transit systems can give a good estimate of length of queues and corresponding waiting time.

Reisman, et al.[81] gave a solution to count crowd from a moving platform that analyze the optic flow generated from the moving objects and moving platform up to a distance of 70m. It is one of the many solutions present in real-time surveillances. Due to the growing demand in the animation and films, new techniques has been developed for creating human crowd scenes which can only be interpreted by previously training data, as crowd counting is greatly dependent on it[82].

Solution using fixed cameras that use standard image-processing techniques can be separated into two types as given in [15]. The first which uses an overhead camera and the second is counting people using people detection and crowd segmentation. Overhead cameras fulfill the limitation of side camera i.e occlusion but it is itself prone to tracking errors as single camera is not enough for tracking and multiple cameras make it more complex. In crowd segmentation, skin color, face detection, motion and other silhouettes are used [83] [84], but all these techniques rely on the quality of the cameras.

Tragically, mass gatherings such as music festivals, sports events or pilgrimage quite often end in terrible crowd disasters with many victims. This gives the motivation to the researchers to find out not only the count but nature and behavior of the crowd. It helps not only to deal with the accidents but to control the situation before it gets too far.

The emergence and dispersal of stationary crowd groups causing the dynamic variations of traffic patterns.[93]

Yang Cong, et al. [85] gave a method to detect abnormal event detection in crowded scenes using sparse representation. Whether a testing sample is abnormal or not is determined by its sparse reconstruction cost, through a weighted linear reconstruction of the over-completed normal bases. Their selection method supports a robust estimation of the situations with minimal size.

An anomaly detection approach was developed for monitoring entrances at installations [86] to take into account the dynamic threat level that often is associated with security applications. A principle strategy associated with this approach was to involve an operator in adapting the sensitivity of the anomaly detector to the threat level. The operator need not be an algorithm expert, so the adaptation either amounted to identifying tracks as either normal or anomalous, or by suggesting physically based features that would be sensitive to a threatening behavior.

Barbara Krausz [87] presents an automatic system for the detection and early warning of dangerous situations during mass events. It is based on optical flow computations and detects patterns of crowd motion that are characteristic for hazardous congestions. By applying an online change-point detection algorithm, the system is capable of identifying changes in pedestrian flow and thus alarms security personnel to take necessary actions.

Detection of different activities perform during multiple people interactions in crowded areas [94]

Louis, et al. [88] introduced a novel framework for modeling the motion patterns of extremely crowded scenes and detecting unusual events. They represented the rich, non-uniform, localized motions patterns with 3D Gaussian distributions of spatio-temporal gradients. The temporal relationship between local spatio-temporal motion patterns is captured via a distribution-based HMM, and the spatial relationship by a coupled HMM.

On a global level, European Union has taken step to fight terrorism by funding project: SEcuRization KEeps(SERKET)[89] to create methods to analyze crowd behaviors. Tracking moving objects, optical flow and crowd-density measurement is generally used in analysis of crowd behavior methods [90] [91] [92].

2) Face Detection:

Person Identification is one of the most important aspects in surveillance and investigation. To identify a person, various techniques and features are taken into account (e.g. gait, posture, clothes) but face has always been the most reliable feature. Researchers have proposed various solutions to detect faces (single and multiple both), some of them are discussed below:

Zoran et al. [95] gave a face detection approach in neural network based method. A neural network based face detector is trained with three multi-layered back propagation neural networks which take three different face representations as an input. The weighted sum of the results from the three networks should give a reliable judgment on the existence of the face patterns.

A face detection technique using Gabor wavelet networks [133] was given in [96] .In [97], the Bayes classifier was adopted with discriminating feature analysis for frontal face detection.

Liu and Chen [100] the adaptive Hidden Markov Model (HMM) for video-based face recognition. In the training phase, a HMM is created for each individual to learn the statistics and temporal dynamics using the Eigen-face image sequence. During the recognition process, the test sequence is analyzed over time by the HMM corresponding to each subject; its identity is determined by the model providing the highest likelihood.

A novel set-to-set similarity measure, the Matched Background Similarity (MBGS) [98] technique is presented for face detection in YouTube videos. Also a semi-supervised learning algorithms based on harmonic functions is adopted for person identification (face detection) in low-quality web cameras [99].

Classification and comparison of recent face occlusion detection techniques review is given in [101].

3) Human Pose and Gait Estimation:

In transit surveillance applications [15], human pose estimation refers to the pose of the entire human body, and not a pose related to a single body part, such as a head pose, that can be used in various applications. While gait refers to a person's manner of walking. Both these features have attracted researchers for Person Identification and Activity Recognition.

Some of the proposed and experimented solutions are discussed:

On the basis of detecting human skin area, WU Song-Lin et al, [102] did experiments and recognized 8 typical sitting postures using Principle Component Analysis (PCA). Firstly, moving object was detected by background contrast attenuation method. Then, considering the clustered skin area in a fixed region of color space field which has an ellipse-like projection in XY plane, the skin area of moving object was extracted. Finally, the behavior recognition was implemented using PCA on the gray-scale image of skin, and the face motion was analyzed according to the time variation of pixel number in facial skin area.

Zhu et al. [103] proposed method which is based on the analysis of two approaches for head pose estimation from an image sequence in Driving Monitoring, that is, principal component analysis (PCA) and 3D motion estimation. The algorithm performs accurate pose estimation by learning the subject appearance on-line.

A real-time model-based human motion analysis system is developed [104] to track and analyze the motion of the human body parts in a video sequence in terms of Body Definition Parameters (BDPs) and Body Animation Parameters (BAPs) for MPEG IV video encoder. The approach contains three phases, segmentation, BDPs generation and BAPs estimation.

A method was introduced in [105] to incorporate both appearance and geometric affine invariances to represent gait sequences. Variant representations were extracted from images and their temporal behavior was modeled. A frame work was modeled such that invariant representations could be modeled and compared in the space of linear dynamical systems.

Sourabh A. Niyogi et al. [106] suggested an algorithm such that a person walking fronto-parallel to the image plane generates a characteristic "braided" pattern in a spatiotemporal (XYT) volume. This algorithm detects this pattern, and fits it with a set of spatiotemporal snakes. The snakes are used to find the bounding contours of the walker. Individual gaits are recognized by applying standard pattern recognition techniques to the contour signals.

4) Loitering:

Loitering is defined as presence of individual or individuals in an area for a period of time longer than the usual (threshold time) [15]. Loitering is of special interest to public surveillance systems since it is common practice of drug dealers, beggars, muggers, grafitti vandals, and among others [a].Though solid work in detection of loitering has not been done so far, some researchers has given some efficient methods.

A person standing near wall for considerable amount of time, but he is not supposed to. Loitering detected [130]

Thi Thi Zin, et al. [107] proposed and investigated a Markov random walk model that can robustly detect loitering individuals in any outdoor public place. The system is divided into three main components. The segmentation module processes the video and extracts images of individual pedestrians by using background subtraction and blob tracking. These images, along with their most probable backgrounds, time stamp, and tracking number are then passed to the feature extraction module proceeding towards the Markov random walk modeling module for detecting loitering people. To determine if a given class is loitering, the time stamps associated with it are analyzed by using stationary and boundary crossing probabilities.

Another method similar to the one given just discussed is given in [108], except for the model for interpretation. A single camera system detects loitering individuals at an inner-city bus stop, it can be used to detect loitering people in any outdoor public place. In spite of Markov random model in [107], the images, along with their most probable backgrounds, time stamp, and tracking number are then passed to the correlation module. The correlation module implements an appearance-based online classification technique that uses the short-term biometric of clothing color to cluster incoming pedestrian images into classes comprised of images of a single pedestrian. To determine if a given class is loitering, the time stamps associated with it are analyzed.

Ye Zhangi et al. [109] present a novel method for judging irregular behavior based on treading track. The background subtraction method is used first to detect moving body, and then it is judged whether someone is suspicious or not on the basis of treading track. Treading track of type spiral or curved are flagged suspicious.

5) Trespassing or Intrusion:

Intrusion or trespassing is defined as the presence of people in a forbidden/NOT-ALLOWED area. A forbidden area can also be defined in terms of temporal or spatial relationships (e.g., time or pedestrian walk). The most popular way of detecting intrusion is trip wire. A trip wire [15] is typically a line drawn over the image, which separates regions into "allow" and "do not allow" areas. Intrusion detection is necessary to detect suicidal behaviors, such as people jumping on the train tracks and obviously for safety purposes e.g. someone going into a health hazardous place. Trespasser hiding [110] can be defined as a blob disappearing in many consecutive frames and the centroid of the blob over time is not constant. Time-dependent HMMs [111] has been successfully used for intrusion violation detection.

6) Miscellaneous:

Alg Saglam et al. [112] proposed new algorithms to detect three types of camera tampering which are camera defocusing, camera movement and covering of a camera view. These adaptive background estimation technique based algorithms uses threshold mechanism (thresholds updated in real-time) to identify moving objects. In terms of camera defocusing, Fourier transform is applied to the images and then a Gaussian window is used to discriminate high frequency components in current image from the low frequencies which is compared with the threshold background. For camera moving, a method is proposed which compares the background image to a delayed background image and the number of different pixels is used to determine whether the camera is moved to a different direction. For camera covering, histograms of current, background and difference image is checked. The difference image is obtained by taking absolute difference between current and background image

A novel Switching Hidden Semi Markov model was presented in [113] which can learn what an occupant normally do during the day from non-segmented training data. Switching Hidden Semi- Markov Model (S-HSMM) introduced in [114] is a special case of the hierarchical model with only two layers. The top layer is a Markov sequence of switching variables, while the bottom layer is a sequence of concatenated HSMMs whose parameters are determined by the switching variable at the top. Thus, the dynamics and duration parameters of the HSMM at the bottom layer are not time invariant, but are "switched" from time to time. S-HSMM outperforms existing HMM models and confirms that both hierarchy and duration information are needed to build accurate activity models in the home and offices..

Elden et al. [114] presents a system that detects human climbing fences. An extended star-skeleton representation consisting of the highest body contour and centroid of the whole body is taken as two stars. Distances between stars and contour points (of other body parts) are computed and smoothed to detect local maximum points. To analyze the resulting time series, a block based discrete Hidden Markov Model (HMM) is built with predefined states.

Person trying to climb fence (illegally) is detected and marked with red rectangular box [132]

Jasper et al. [115] presents a method for automatically detecting and recognizing unusual events on stairs from video data. Optical flow features\ and a hidden Markov model is used to detect unusual events in data.

B. HEALTH CARE:

Processing and semantically deriving information from videos has aid Health and Care sector of the society in recent times. From sleeping disorders to respiratory and care for elderly people, researchers are finding novel methods to make the society with respect to health a better place. Though

comparatively to others, huge amount of work has been done in this field, we will look up to some of the recent works.

Patrick et al. [116] presented a novel method for learning normal patterns of behavior, intended to support home-based care of vulnerable elderly people. The proposed method is based on building a spatial map of normal inactivity, from a set of training data. This map represents inactivity regions as a 2D Mixture of Gaussians model, learned using EM. Spatial MoG model is used to build a pair of HMMs: the first representing normal sequences of inactivity, and the other representing arbitrary behavior. The advantage of using this method is it is not dependent on identifying specific activities, or modeling time-series data.

Monika et al. [117] gave a method for image based human activity recognition, in a smart environment. They used background subtraction and skeletisation as image processing techniques, combined with Artificial Neural Networks for human posture classification and Hidden Markov Models for activity interpretation. By this approach basic human action such as walking, rotating, sitting and bending up/down, lying and falling were recognized. The method can be applied in smart houses, for elderly people who live alone.

Just like the above two methods, TIMC-IMAG laboratory [118] developed Health Smart Homes called "HIS". These smart Homes are composed of several sensors to monitor the activities of daily living of the patients.

While learned models of spatial context are used in conjunction with a tracker to achieve fall detection for older people in [119].

An algorithm for dining activity analysis in a nursing home is given in [120]. Based on several features, including motion vectors and distance between moving regions in the subspace of an individual person, a Hidden Markov model is proposed to characterize different stages in dining activities with certain temporal order. Using HMM model, the start (and ending) of individual dining events with high accuracy and low false positive rate are identified. This approach could be successful in assisting caregivers in assessments of resident's activity levels over time.

Between 1993 and 1995, Lionel [121] et al. collaborated with the Medical Systems Division of Oxford Instruments to develop a commercial version of a neural network system which detect disorder of 'Sleep Walking' and classify it, known as QUESTAR (QUantification of EEG and Sleep Technologies Analysis and Review). Both supervised and unsupervised methods were taken for 2D visualization for the sleep which are categorized into: weakness, light sleep and deep sleep, responsible for the disorder.

C. SPORTS:

Sports broadcasts [122] constitute a major percentage in the total of public and commercial television broadcasts. The growing demands of the consumers/viewer require advances in video capturing and video processing ability. With the availability of large storage capabilities and more TV channels with full coverage of large sport events, the organization and search in this field becomes more appealing. This situation has attracted many researchers in finding out the interesting segments in the whole video with respect to the perspective of the consumers/viewers. Other growing demands of viewers are new enhancement and presentation techniques that provide a better viewing experience.

The video abstract is a very important tool for sports video, as several periods during some games may be boring to the consumers and watching the abstract can save a lot of time. Several researchers have used audio features to detect the highlights, because audio contains information about audience applause and of the excitement in the voice of the commentator. Rui et al. [125] used audio features such as excited speech/ high pitch and baseball hits to detect the highlights. Pan et al. [123] describes a method by localizing the semantically important events by detecting the slow motion replays in sports and also filters the commercial slow motion replays. Hidden Markov Model (HMM) combined with viterbi algorithm is used. The advantage of this approach is that the user has more control about what he wants to view and the probability to miss an interesting scene is smaller.

Coaches and players are interested in information extracted from videos for improving their performance and tactics in later games. The consumers/viewers are interested in such results for enjoying sports video with additional statistical information which are generally shown before or after the game ends. The main tasks in tactics and performance analysis are to find the traces and the actions of players. A lot of work has been one in this field as it is largely dependent on tracking and segmentation. Pingali et al. [124] describes in detail a real time tracking system which produces spatio-temporal trajectories of the motion of the player and the ball which can provide useful game related statistics. Detection of tennis strokes are given in [126] while tracking of a particular individual and run time velocity analysis is presented in [127].

Ball Tracking in Decision Review System (DRS) is used in all forms of cricket today [132]

Also, for refree assistance on the field [122], US Open and Australian Open use video systems to display the landing positions that are very close to the boundary lines of the court, and Cricket for LBW decisions and ball tracking, where the technology is provided by Hawkeye [128]. Also in 2014 FIFA World Cup, Goal Line technology [129] has been introduced along with detection of off-sides in play already present.

VII. POSSIBLE FUTURE DIRECTION

Although a large amount of work has been done in human activity and behavior recognition, there are still many open issues needed to be investigated in future, especially in the following areas. In terms of applications, in health care, with the development of smart homes, a step towards a safer and healthier society has begun but they are out of reach due to high monetary investment, hence cheaper and effective hardware and techniques need to be developed in near future so that more people can afford and adopt. There is a lack of platform where the techniques proposed by the scientists and researchers can be compared, i.e. advantages and disadvantages of a recognition technique for a particular context is known but a general comparison framework does not exist. Hence, development of such a method is also a future research topic. Due to the fact that large amount of knowledge is contained in surrounding scene and task context, a same behavior may have different meanings when the surrounding scene and task context changed. However, most existing techniques for human behavior recognition still rely on the assumption of specific circumstance statistics, which at the same time degrades the popularity of recognition technique. It remains an open issue to understand human behavior in complex scenes. In other words, recognition of behavior patterns constructed by self-organizing and self-learning from unknown scenes is a future research direction.

ACKNOWLEDGMENT

The author would like to thank Scientist Mr. Sandip Kumar Maganlal Vaniya, Information Technology Cell, Central Salt and Marine Chemicals Institute, Bhavnagar, Gujrat, India for his guidance, involvement and support in the completion of this paper.

REFRENCES

[1] "A Survey on Behavior Analysis in Video Surveillance Applications" Teddy Ko Raytheon Company, USA
[2] en.wikipedia.org/wiki/Background_subtraction
[3] http://en.wikipedia.org/wiki/Optical_flow
[4] A Bayesian Computer Vision System for Modeling Human Interactions" Nuria M. Oliver, Barbara Rosario, and Alex P. Pentland, IEEE Trans. PAMI, 2000
[5] http://computervision.wikia.com/wiki/Motion_segmentation
[6] "A Survey on Visual Surveillance of Object Motion and Behaviors" Weiming Hu, Tieniu Tan, Liang Wang, and Steve Maybank.
[7] "Recognising Human Behaviours in Complex Environments" Nam Thanh Nguyen
[8] "Adaptive background mixture models for real-time tracking" Chris Stauffer W.E.L Grimson The Artificial Intelligence Laboratory Massachusetts Institute of Technology Cambridge,MA
[9] "Understanding Background Mixture Models for Foreground Segmentation" P. Wayne Power Johann A. Schoonees Industrial Research Limited, Auckland, New Zealand
[10] "A Review on Video-Based Human Activity Recognition" Shian-Ru Ke , Hoang Le Uyen Thuc , Yong-Jin Lee ,Jenq-Neng Hwang , Jang-Hee Yoo , Kyoung-Ho Choi
[11] "Research on GMM Background Modelling and its Covariance Estimation" Journal on Advanced Materials Research (Volumes 383 - 390). Yun Chu Zhang, Ru Min Zhang, Shi Jun Song, (2011)
[12] http://en.wikipedia.org/wiki/Expectation_maximization_algorithm
[13] https://www.youtube.com/watch?v=AnbiNaVp3eQ
[14] "A Survey on Behavior Analysis in Video Surveillance for Homeland Security Applications", Teddy Ko
[15] "Understanding Transit Scenes: A Survey on Human Behavior-Recognition Algorithms", Joshua Candamo, Matthew Shreve, , Dmitry B. Goldgof, Deborah B. Sapper, and Rangachar Kasturi, IEEE TRANSACTIONS ON INTELLIGENT TRANSPORTATION SYSTEMS, VOL. 11, NO. 1, MARCH 2010

[16] "Model of Human Visual-Motion sensing," A. Watson and A. Ahumada, Jr., J. Opt. Soc. Am., A 2, pp. 322-342, 1985.

[17] "Real-time Object Classification in Video Surveillance Based on Appearance Learning", Lun Zhang, Stan Z. Li, Xiaotong Yuan and Shiming Xiang, Center for Biometrics and Security Research & National Laboratory of Pattern Recognition Institute of Automation,Chinese Academy of Science ,Beijing,P.R.China

[18] "Motion Pattern-Based Video Classification and Retrieval", Yu-Fei Ma, EURASIP Journal on Applied Signal Processing 2003:2, 199–208 ©2003 Hindawi Publishing Corpor"Motion-based segmentation and contour-based classification of video objects", Gerald Kühne, Stephan Richter, Markus Beier, University of Mannheim, Mannheim, Germany, 2001

[19] "COMPUTER SYSTEMS THAT LEARN:AN EMPIRICAL STUDY OF THE EFFECT OF NOISE ON THE PERFORMANCE OF THREE CLASSIFICATION METHODS", James R. Nolan.

[20] "Multivariate analysis: Methods and applications". New York: Dillon, W. R. & Goldstein, (1984)

[21] "Neural network design. Boston, Hagan, M. T., Demuth," H. B., & Beale,(1996)

[22] "Introduction to artificial neural systems", Zurada, J. M. St. Paul, MN: West (1992)

[23] "Learning efficient classification procedures and their application to chess end games. Quinlan, J, R, (1983).

[24] "Pfmder:real-time tracking of the human body," C. R. Wren, A. Azarbayejani, T. Darrell, and A. P. Pentland, IEEE Trans. Pattern Anal. Machine Intell., vol. 19, pp.780–785, 1997.

[25] "Human tracking and silhouette extraction for humanrobot interaction systems,"J.-H. Ahn, et al., Pattern Analysis & Applications, vol. 12

[26] "Robust tracking of position and velocity with Kalman snakes," N. Peterfreund, IEEE Transactions on Pattern Analysis and Machine Intelligence, vol. 21

[27] "Color-Based Probabilistic Tracking," P. Perez, et al., in ECCV, 2002

[28] "Ribbon-based motion analysis of human body movements," I. C. Chang and H. Chung-Lin, in Proceedings of the 13th International Conference on Pattern Recognition, 1996

[29] "Tracking loose-limbed people," L. Sigal, et al., in Computer Vision and Pattern Recognition, ,2004

[30] "A MODEL-BASED GAZE TRACKING SYSTEM", RAINER STIEFELHAGEN, JIE YANG and ALEX WAIBEL, Interactive System Laboratories Carnegie Mellon University, USA and University of Karlsruhe, Germany

[31] "Understanding and Applying Kalman Filtering", Lindsay Kleeman. Department of Electrical and Computer Systems Engineering, Monash University, Clayton

[32] "Kalman Filter Based Tracking in an Video Surveillance System," Caius, Cristina, Florin, 10th International Conference on DEVELOPMENT AND APPLICATION SYSTEMS, Suceava, Romania, May 27-29, 2010

[33] http://www.vision.caltech.edu/koller/ModelTracking.html

[34] http://en.wikipedia.org/wiki/Condensation_algorithm

[35] "Conditional density propagation for visual tracking," M. Isard and A. Blake, Int. J. Computer Vision 29(1),1998, pp. 4-28

[36] http://en.wikipedia.org/wiki/Dynamic_Bayesian_network

[37] "BAYESIAN DYNAMIC MODELLING", Mike West, Department of Statistical Science, Duke University

[38] "Motion-Based Recognition: A Survey," Cedras, C. & Shah, M. (1995). Image and Vision Computing

[39] "A Survey of Vision-Based Trajectory Learning and Analysis for Surveillance," Morris, B. & Trivedi, M. (2008). IEEE Trans. on Circuits and Systems for Video Technology

[40] "Learning Semantic Scene Models From Observing Activity in Visual Surveillance," Makris, D. & Ellis, T. (2005). , IEEE Transactions on Systems, Man, and Cybernetics – Part B: Cybernetics, Vol. 35, No. 3, pp. 397- 408.

[41] "Tracking video objects in cluttered background," Cavallaro, A., Steiger, O. & Ebrahimi, T. (2005). IEEE Transactions on Circuits and Systems for Video Technology

[42] "Neural Network Based Threat Assessment for Automated Visual Surveillance," Jan, T. (2004). in Proceedings of IEEE International Joint Conference on Neural Networks

[43] "The Shanghai-Hongkong Team at MediaEval2012: Violent Scene Detection Using Trajectory-based Features", Yu-Gang Jiang, Qi Dai, Chun Chet Tan, Xiangyang Xue Chong-Wah Ngo

[44] "Video-Based Abnormal Human Behavior Recognition—A Review", Oluwatoyin P. Popoola, and Kejun Wang

[45] "Anomaly detection:A survey", V. Chandola, A. Banerjee, andV.Kumar, ACM Comput. Surveys, vol. 41, pp. 1–58, Jul. 2009.

[46] "Everything you know about Dynamic Time Warping is Wrong", Chotirat Ann, Ratanamahatana Eamonn Keogh, Department of Computer Science and Engineering University of California, Riverside

[47] "Human action recognition using Dynamic Time Warping", Sempena, S. ; Sch. of Electr. Inf. Eng., Inst. Teknol. Bandung, Bandung, Indonesia ; Maulidevi, N.U. ; Aryan, P.R. Electrical Engineering and Informatics (ICEEI), 2011 International Conference

[48] "IMAGE TEMPLATE MATCHING ON MIMD HYPERCUBE MULTICOMPUTERS", Sanjay Ranka and Sartaj Sahni, University of Minnesota

[49] "A state-based technique to the representation and recognition of gesture," A. F. Bobick and A. D.Wilson, IEEE Trans. Pattern Anal. Machine Intell., vol. 19, pp. 1325–1337, Dec. 1997.

[50] "Recognition of dexterous manipulations from time varying images,"K. Takahashi, S. Seki, H.Kojima, and R. Oka, in Proc. IEEEWorkshop Motion of Non-Rigid and Articulated Objects, Austin, TX, 1994, pp. 23–28.

[51] "Understanding video events: a survey of methods for automatic interpretation of semantic occurrences in video," G. Lavee, E. Rivlin, and M. Rudzsky, IEEE TRANSACTIONS ON SYSTEMS, MAN, AND CYBERNETICS—PART C, vol. 39, pp. 489–504, Sept. 2009.

[52] "Temporal classification of natural gesture and application to video coding," A. D. Wilson, A. F. Bobick, and J. Cassell, in Proc. IEEE Conf. Computer Vision and Pattern Recognition, 1997, pp. 948–954.

[53] "DETECTING HUMAN ACTIVITIES IN RETAIL SURVEILLANCE USING HIERARCHICAL FINITE STATE MACHINE", Hoang Trinh, Quanfu Fan, Jiyan Pan, Prasad Gabbur, Sachiko Miyazawa, Sharath Pankanti

[54] "HUMAN ACTIVITY RECOGNITION FROM BASIC ACTIONS USING FINITE STATE MACHINE", Nattapon Noorit, and Nikom Suvonvorn

[55] "Discovery and Segmentation of Activities in Video," M. Brand and V. Kettnaker, IEEE Transactions on Pattern Analysis and Machine Intelligence, Vol. 22, No. 8, pp. 844-851, 2000.

[56] "Automatic Understanding of Human Behavior in Videos: A review" Mourad Bouzegza , M. Elarbi-Boudihir College of Computer Science1, College of Science2. Imam University, Riyadh, Kingdom of Saudi Arabia

[57] "Visual Recognition of American Sign Language Using Hidden Markov Models", Thad Starner and Alex Pentland

[58] "Coupled hidden Markov models for complex action recognition," Matthew Brand, Nuria Oliver, and Alex Pentland

[59] "Dynamic bayesian networks: Representation, inference and learning," K. P. Murphy, 2002.

[60] "A hierarchical bayesian network for event recognition of human actions and interactions," S. Park, in Association For Computing Machinery Multimedia Systems Journal, pp. 164–179, 2004.

[61] "Artificial Neural Networks for Beginners", Carlos Gershenson

[62] "Artificial Neural Networks", Girish Kumar Jha, IARI, PUSA, New Delhi

[63] "Neural Networks: A Systematic Introduction", Ra'ul Rojas

[64] "An Introduction to Neural Networks", Vincent Cheung, Kevin Cannons

[65] "Artifical Neural Networks: A Tutorial", Anil K. Jain, Michigan State University

[66] "Machine Learning: Multi Layer Perceptrons", Prof. Dr. Martin Riedmiller

[67] "Radial Basis Function Networks: Introduction, Introduction to Neural Networks : Lecture 12" © John A. Bullinaria, 2004

[68] "Neural Network Based Video Surveillance System", Alberto Amato1, Vincenzo Di Lecce1, Vincenzo Piuri2

[69] "Recognition of visual activities and interactions by stochastic parsing," Y. Ivanov and A. Bobick, IEEE Transactions on Pattern Analysis and Machine Intelligence, vol. 22, pp. 852–872, 2000.

[70] "Multi-object behavior recognition by event driven selective attention method," T.Wada and T. Matsuyama, IEEE Trans. Pattern Anal. Machine Intell., vol. 22, pp. 873–887, Aug. 2000.

[71] "Understanding video events: a survey of methods for automatic interpretation of semantic occurrences in video," G. Lavee, E. Rivlin, and M. Rudzsky, IEEE TRANSACTIONS ON SYSTEMS, MAN, AND CYBERNETICS—PART C, vol. 39, pp. 489–504, Sept. 2009.

[72] "Understanding manipulation in video", M Brand

[73] "An Efficient Probabilistic Context-Free Parsing Algorithm that Computes Prefix Probabilities", Andreas Stolcke

[74] "Early Parsing for 2D Stochastic Context Free Grammars", Andelo Martinovic and Luc Van Gool

[75] en.wikipedia.org/wiki/A_priori_and_a_posteriori

[76] "Carl Adam Petri and "Petri Nets"", Wilfried Brauer, Wolfgang Reisig

[77] "Modeling and recognition of video events with Fuzzy Semantic Petri Nets", Piotr Szwed, AGH University of Science and Technology

[78] http://en.wikipedia.org/wiki/Time_delay_neural_network

[79] "Fast pattern matching with time-delay neural networks", Heiko Hoffmann, Michael D. Howard, and Michael J. Daily

[80] "Crowd detection in video sequences," P. Reisman, O. Mano, S. Avidan, and A. Shashua, in Proc. Intell. Vehicles Symp., 2004, pp. 66–71.

[81] "On crowd density estimation for surveillance", H. Rahmalan, M. S. Nixon, and J. N. Carter,

[82] "Fast crowd segmentation using shape indexing," L. Dong, V. Parameswaran, V. Ramesh, and I. Zoghlami, in Proc. IEEE Int. Conf. Comput. Vis., 2007, pp. 1–8.

[83] "Tracking and segmenting people with occlusions by a sample consensus based method," H.Wang and D. Suter, in Proc. IEEE Int. Conf. Image Process., 2005, vol. 2, pp. 410–413.

[84] "Abnormal Event Detection in Crowded Scenes using Sparse Representation", Yang Cong, Junsong Yuan and Ji Liu, State Key Laboratory of Robotics, Shenyang Institute of Automation, Chinese Academy of Sciences, China

[85] "An Approach to Detecting Crowd Anomalies for Entrance and Checkpoint Security", Holly Zelnio, B.S. Wright State University 2010

[86] "Automatic detection of dangerous motion behavior in human crowds", Barbara Krausz, Christian Bauckhage, 2011 8th IEEE International Conference on Advanced Video and Signal Based Surveillance

[87] "Anomaly Detection in Extremely Crowded Scenes Using Spatio-Temporal Motion Pattern Models", Louis Kratz, Ko Nishino ,Department of Computer Science Drexel University

[88] "Intelligent environments for problem solving by autonomous systems," S. Antipolis, Institut National de Recherche en Informatique et en Automatique, Rocquencourt, France, p. 41, 2007.

[89]   "Counting crowded moving objects," V. Rabaud and S. Belongie, in Proc. IEEE Int. Conf. Comput. Vis. Pattern Recog., 2006, vol. 1, pp. 705–711.

[90]   "Hidden Markov models for optical flow analysis in crowds," E. L. Andrade, S. Blunsden, and R. B. Fisher, in Proc. Int. Conf. Pattern Recog., 2006, vol. 1, pp. 460–463.

[91]   "Image processing techniques for crowd density estimation using a reference image," J. H. Yin, S. A. Velastin, and A. C. Davies, in Proc. Asian Conf. Comput. Vis., 1995, pp. 489–498.

[92]   http://crowdbehavior.org/

[93]   http://rose.ntu.edu.sg/research/Objec%20Search/Pages/Anomaly-Detection.aspx

[94]   "FACE DETECTION APPROACH IN NEURAL NETWORK BASED METHOD FOR VIDEO SURVEILLANCE", Zoran Bojkovic,, Andreja Samcovic

[95]   "A Framework for Face Recognition from Video Sequences Using GWN and Eigenfeature Selection", Teófilo Emídio de Campos, Rogério Schmidt Feris, Roberto Marcondes Cesar Junior

[96]   "A bayesian discriminating features method for face detection", Liu, C. IEEE Transactions on Pattern Analysis and Machine Intelligence 25(6), 725–740 (2003)

[97]   "Face Recognition in Unconstrained Videos with Matched Background Similarity", Lior Wolf, Tal Hassner, Itay Maoz,

[98]   "Person Identification in Webcam Images: An Application of Semi-Supervised Learning" Maria-Florina Balcan, Avrim Blum, Patrick Pakyan Choi, John Lafferty, Brian Pantano , Mugizi Robert , Xiaojin Zhu

[99]   "Video-based face recognition using adaptive hidden Markov models", Xiaoming Liu, Tsuhan Chen, IEEE Computer Vision and Pattern Recognition, 2003.

[100] "A Survey of Face Occlusion Detection for Visual Surveillance System", Theekapun Charoenpong

[101] "Human Behavior Recognition Based on Sitting Postures", WU Song-Lin, CUI Rong-Yi, Intelligent Information Processing Lab., Department of Computer Science & Technology, Yanbian University, Yanji, China

[102] "Head pose estimation for driver monitoring," Y. Zhu and K. Fujimura, in Proc. IEEE Intell. Vehicles Symp., 2004, pp. 501–506.

[103] "Model-Based Human Body Motion Analysis for MPEG IV Video Encoder", Chung-Lin Huang and Chung-Chin Lin, Electrical Engieering Department, National Tsing-Hua University, Hsin-Chu, Taiwan\

[104] "Gait Recognition using Dynamic Affine Invariants", Alessandro Bissacco, Payam Saisan, Stefano Soatto

[105] "Analyzing and Recognizing Walking Figures in XYT", Sourabh A. Niyogi

[106] "A Markov Random Walk Model for Loitering People Detection", Thi Thi Zin, Pyke Tin and Takashi Toriu, Hiromitsu Hama, 2010 Sixth International Conference on Intelligent Information Hiding and Multimedia Signal Processing

[107] "Detection of Loitering Individuals in Public Transportation Areas", Nathaniel D. Bird, Osama Masoud, Nikolaos P. Papanikolopoulos and Aaron Isaacs, IEEE TRANSACTIONS ON INTELLIGENT TRANSPORTATION SYSTEMS, VOL. 6, NO. 2, JUNE 2005

[108] "IRREGULAR BEHAVIOR RECOGNITION BASED ON TREADING TRACK", YE ZHANG1, ZHI-JING LIU2, Proceedings of the 2007 International Conference on Wavelet Analysis and Pattern Recognition, Beijing, China, 2-4 Nov. 2007

[109] "Tracking-based event detection for CCTV systems," L. M. Fuentes and S. A. Velastin

[110] "Time-dependent HMMs for visual intrusion detection," V. Kettnaker, in Proc. IEEE Int. Conf. Comput. Vis. Pattern Recog. Workshop, 2003, vol. 4, p. 34.

[111] "ADAPTIVE CAMERA TAMPER DETECTION FOR VIDEO SURVEILLANCE", ALG SAGLAM, 2009

[112] "Activity Recognition and Abnormality Detection with the Switching Hidden Semi-Markov Model", Thi V. Duong, Hung H. Bui, Dinh Q. Phung, Svetha Venkatesh

[113] "Detection of Fence Climbing from Monocular Video", Elden Yu and J.K. Aggarwal

[114] "Automated Detection of Unusual Events on Stairs", Jasper Snoek Jesse Hoey Liam Stewart Richard S. Zemel

[115] "Using Inactivity to Detect Unusual behavior", Patrick Dickinson and Andrew Hunter

[116] "Human Activity Recognition in Smart Environments", Monica-Andreea Dragan   and Irina Mocanuy

[117] "Identification of inactivity behavior in Smart Home", J. Poujaud, N. Noury, senior member, IEEE, 30th Annual International IEEE EMBS Conference Vancouver, British Columbia, Canada, August 20-24, 2008

[118] "Activity Summarisation and Fall Detection in a Supportive Home Environment", Hmmadi Nait-Charif and Stephen J. McKenna

[119] "Dining Activity Analysis Using a Hidden Markov Model", Jiang Gao1, Alexander G. Hauptmann1, Ashok Bharucha2, and Howard D. Wactlar1, 17th International Conference on Pattern Recognition (ICPR'04), Cambridge, United Kingdom, August 23-26, 2004

[120] "Neural network analysis of sleep disorders", Lionel Tarassenko, Mayela Zamora and James Pardey

[121] "Current and Emerging Topics in Sports Video Processing", Xinguo Yu, Dirk Farin

[122] "Detection of slow-motion replay segments", Pan, H., P. Van Beek and M.I. Sezan, 2001. Proceedings of IEEE International Conference on Acoustics, Speech and Signal Processing, (ICASSP '01), Illinois Univ., Urbana, IL, 3: 1649-1652.

[123] "Real timetracking for tennis broadcasts", Pingali, G.S., Y. Jean and I. Carlbom, 1998. Proceedings of the IEEE Computer Society Conference on Computer Vision and Pattern Recognition, CVPR '98, Washington, DC, USA, pp: 260.

[124] "Automatically extracting highlights for TV Baseball programs", Y. Rui, A. Gupta, A. Acero, ACM MM00, pp105-115, Oct. 2000

[125] "Human activity detection in sports video", Takehito Ogata

[126] "Human Object Detection and Tracking using Background Subtraction for Sports Applications", R.Manikandan, R.Ramakrishnan, Tamil Nadu Physical Education and Sports University, Chennai, India

[127] http://www.hawkeyeinnovations.co.uk.

[128] http://eurekaweb.fr/wp/goal-line-technology-en/

[129] www.securiton.de

[130] www.dailymail.co.uk

[131] www.sstpgroup.co.uk

[132] "Efficient Head Pose Estimation with Gabor Wavelet Networks", Volker Kr¨uger, Sven Bruns and Gerald Sommer, Computer Science Institute, Christian-Albrechts   University